# THE POETRY OF RUBIDIUM

# The Poetry of Rubidium

Walter the Educator™

Silent King Books a WhichHead Imprint

Copyright © 2023 by Walter the Educator™

All rights reserved. No part of this book may be reproduced in any manner whatsoever without written permission except in the case of brief quotations embodied in critical articles and reviews.

First Printing, 2023

Disclaimer
This book is a literary work; poems are not about specific persons, locations, situations, and/or circumstances unless mentioned in a historical context. This book is for entertainment and informational purposes only. The author and publisher offer this information without warranties expressed or implied. No matter the grounds, neither the author nor the publisher will be accountable for any losses, injuries, or other damages caused by the reader's use of this book. The use of this book acknowledges an understanding and acceptance of this disclaimer.

"Earning a degree in chemistry changed my life!"
- Walter the Educator

dedicated to all the chemistry lovers, like myself, across the world

# CONTENTS

Dedication . . . . . . . . . . v

Why I Created This Book? . . . . . . . . . 1

**One** - Element Of Affection . . . . . . . . 2

**Two** - Captivating Hue . . . . . . . . . 4

**Three** - Beyond Compare . . . . . . . . 6

**Four** - Cosmic Wings . . . . . . . . . 8

**Five** - Forever To Be . . . . . . . . . 10

**Six** - Wonders Amiss . . . . . . . . . 12

**Seven** - Revealing The Unknown . . . . . . . 14

**Eight** - Marvel Divine . . . . . . . . . 16

**Nine** - Universe Gleams . . . . . . . . 18

**Ten** - Rubidium's Presence . . . . . . . . 20

**Eleven** - Atomic Embrace . . . . . . . . 22

**Twelve** - Time And Space . . . . . . . . 24

| | | |
|---|---|---|
| **Thirteen** - Rubidium's Grace | . . . . . . . . | 26 |
| **Fourteen** - Infinite Reflection | . . . . . . . . | 28 |
| **Fifteen** - Timeless Themes | . . . . . . . . | 30 |
| **Sixteen** - Flickering Flame | . . . . . . . . | 32 |
| **Seventeen** - Element Rare | . . . . . . . . | 34 |
| **Eighteen** - Day By Day | . . . . . . . . | 36 |
| **Nineteen** - Spectral Line | . . . . . . . . | 38 |
| **Twenty** - Mesmerizing Sight | . . . . . . . . | 40 |
| **Twenty-One** - Celestial Zeal | . . . . . . . . | 42 |
| **Twenty-Two** - Vibrant Sensation | . . . . . . . . | 44 |
| **Twenty-Three** - Forever Unite | . . . . . . . . | 46 |
| **Twenty-Four** - Infinitely Kind | . . . . . . . . | 48 |
| **Twenty-Five** - Science Strives | . . . . . . . . | 50 |
| **Twenty-Six** - Silently Speaks | . . . . . . . . | 52 |
| **Twenty-Seven** - Number 37 | . . . . . . . . | 54 |
| **Twenty-Eight** - Flames Of Curiosity | . . . . . . . . | 56 |
| **Twenty-Nine** - Wide And Raw | . . . . . . . . | 58 |
| **Thirty** - Honor Your Name | . . . . . . . . | 60 |
| **Thirty-One** - Rubidium's Touch | . . . . . . . . | 62 |
| **Thirty-Two** - Mineral Found | . . . . . . . . | 64 |

**Thirty-Three** - Catalyst For Knowledge . . . 66

**Thirty-Four** - Spectacle To View . . . . . . . 68

**Thirty-Five** - The Starlight's Secret . . . . . . 70

About The Author . . . . . . . . . . . . . . 72

# WHY I CREATED THIS BOOK?

Creating a poetry book about the chemical element of Rubidium was an intriguing and unique endeavor. Rubidium, with its atomic number 37 and symbol Rb, possesses certain qualities that can be metaphorically explored through poetry. It is a soft, silvery-white alkali metal with a rich history and fascinating properties. By delving into its characteristics, such as its reactivity, conductivity, and use in scientific research, I can draw parallels to human emotions, relationships, and the world around us. This fusion of science and art result in a captivating and thought-provoking collection of poems that adds a new dimension to both subjects.

# ONE

# ELEMENT OF AFFECTION

In the realm of elements, where secrets do reside,
A shimmering presence, Rubidium does confide.
A gem of the periodic table, shining so bright,
A tale of allure, a dance in the starry night.

Oh, Rubidium, with your atomic embrace,
A flame that burns with elegance and grace.
Your aura, like moonlight on a calm sea,
Captivates the senses, sets the spirit free.

With an atomic number of thirty-seven,
You hold secrets that defy heaven.
Your electrons dance in a delicate waltz,
Guided by forces, their movements exalts.

From the depths of Earth, where you rest,
A treasure to be discovered, a gem to be blessed.

Your presence ignites the laboratory's air,
A catalyst of wonders, beyond compare.
    In nature's embrace, you find your place,
In rocks and minerals, a celestial trace.
From stars to the earth, a cosmic connection,
Rubidium, the element of affection.
    Oh, Rubidium, you enchant with your hue,
A crimson flame, a tale that rings true.
In the realm of elements, you hold the key,
To mysteries untold, for all to see.

# TWO

# CAPTIVATING HUE

In the realm of elements, a treasure lies,
A shimmering presence with captivating eyes.
Rubidium, a gem of the periodic table,
Elegant and graceful, it is able.

With atomic number thirty-seven it is known,
A secret keeper, a mystery to be shown.
Its hue, a hue of enchantment and allure,
A catalyst of wonders, it does ensure.

Nature's embrace, where it finds its place,
In rocks and minerals, it leaves a trace.
A cosmic connection, beyond our reach,
In stars it dwells, the heavens it does teach.

Oh Rubidium, a jewel of the unseen,
A dance of electrons, a majestic sheen.
Unveiling the secrets, you hold within,
Unlocking the mysteries, where wonders begin.

In laboratories, your secrets we explore,
Your presence inspires, forevermore.
With each discovery, the world expands,
A testament to the power in your hands.

Oh Rubidium, with your captivating hue,
You light up the darkness, shining through.
A gem of the elements, a treasure untold,
In your elegance and grace, we behold.

# THREE

# BEYOND COMPARE

In the realm of elements, a gem does reside,
A shimmering presence, with elegance and pride.
Rubidium, they call it, a treasure to behold,
A catalyst for wonders, a story yet untold.

Oh Rubidium, with eyes so captivating,
A secret keeper, a mystery captivating.
In laboratories it dances, unlocking unseen,
Revealing hidden truths, where wonders convene.

A cosmic connection, to stars and the heavens,
A bridge between worlds, where knowledge is leavened.
It whispers of secrets, untold and unseen,
Inspiring the curious, the seekers, the keen.

Oh Rubidium, your hue is so captivating,
A vibrant red flame, in darkness undulating.

You light up the shadows, with fiery grace,
A gem among elements, a treasure in this space.
   So let us raise a toast, to Rubidium so rare,
A substance of wonders, beyond compare.
Through laboratories it dances, unlocking new doors,
Revealing the mysteries, that nature adores.
   Oh Rubidium, you are a gem, a treasure untold,
With elegance and grace, your story unfolds.
In laboratories you ignite, the air with your flare,
A catalyst for wonders, beyond compare.

# FOUR

## COSMIC WINGS

In the realm of elements, a shimmering presence,
Rubidium emerges, a catalyst for wonders,
Its atomic dance, a symphony of essence,
Unveiling secrets, as the night sky thunders.

From the depths of nature, it gracefully arose,
A flame in the darkness, a rhapsody of light,
With a ruby hue, its beauty it bestows,
A harbinger of brilliance, a celestial sight.

Oh Rubidium, captivator of hearts,
Unveiling truths, with your mystical art,
You unlock the mysteries, where knowledge imparts,
A radiant gem, in science's chart.

In its atomic core, secrets reside,
A key to the cosmos, a cosmic guide,

Through the spectral lines, it whispers and sings,
Revealing the universe, on its cosmic wings.
   Oh Rubidium, in your elegance we find,
A catalyst for wonders, a treasure to behold,
With grace and charm, you leave us entwined,
A gem among elements, a story yet untold.

# FIVE

## FOREVER TO BE

In the realm of elements, a gem I see,
Rubidium, radiant and free.
With atomic embrace, it beckons me near,
A flame that burns with elegance and cheer.

In the depths of the periodic table it lies,
A treasure hidden from prying eyes.
Its vibrant hues, a captivating sight,
A dance of electrons, shimmering in the light.

A secret it keeps, in its atomic core,
Whispering tales of the cosmic lore.
For in rocks and minerals, it finds its home,
A cosmic connection, to the stars it roams.

From stellar explosions, it was born,
A celestial journey, its essence worn.
It floats through the cosmos, a celestial tide,
A cosmic companion, on this wondrous ride.

Rubidium, a catalyst for wonders untold,
A bridge between worlds, both new and old.
With each interaction, it weaves a thread,
Connecting the universe, with elegance and stead.

So let us celebrate this radiant gem,
Rubidium, a presence that won't condemn.
A spark of creation, in the cosmic sea,
A gem among elements, forever to be.

# SIX

# WONDERS AMISS

In the realm of elements, a gem does reside,
A fiery spirit, with secrets to confide.
Rubidium, the name that graces its fame,
A catalyst for wonders, a burning flame.

With an atomic number of thirty-seven,
Its presence on Earth, a cosmic heaven.
Its electrons dance, in orbits they spin,
Creating a symphony, a celestial din.

In laboratories, it reveals hidden truths,
Unlocks mysteries, like ancient sooths.
Its spectral lines, a language untold,
A key to the universe, a tale to unfold.

In the heart of stars, it finds its birth,
A cosmic connection, a bridge to traverse.

From supernovae's fiery demise,
Rubidium emerges, a celestial prize.
   Through the vast expanse, it gracefully floats,
Its vibrant hues, like celestial notes.
A secret keeper, it whispers in space,
Holding the secrets of the cosmic embrace.
   Rubidium, a marvel, so rare and divine,
A story untold, in its atomic design.
From laboratories to the cosmic abyss,
Its presence reminds us, of wonders amiss.

# SEVEN

# REVEALING THE UNKNOWN

In the realm of atoms, a gem unseen,
Lies a treasure in shades of crimson sheen.
Rubidium, noble and rare,
In the dance of electrons, a cosmic affair.

Within its core, secrets untold,
A catalyst of wonders, a story to unfold.
Its valiant electrons, they sway and spin,
Creating a symphony, a harmony within.

In the vast expanse of the atomic dance,
Rubidium's elegance, a captivating trance.
With each leap and bound, it lights the way,
Guiding us through night, as stars at play.

From laboratories to celestial skies,
Rubidium's presence, a cosmic surprise.

Its atomic beauty, a celestial flame,
Burning bright, with no one to blame.
    Oh, Rubidium, mysterious and grand,
A cosmic connection, we can't understand.
So let us marvel, and let us explore,
The wonders you hold, forevermore.
    In the depths of the universe, you shine,
A priceless gem, a treasure of time.
Unlocking secrets, revealing the unknown,
Rubidium, a muse to call our own.

# EIGHT

## MARVEL DIVINE

In cosmic dance, where stars collide,
A wondrous element does reside.
Rubidium, a bridge between the worlds,
In its presence, secrets unfurl.
    Unlocking hidden truths it holds,
A catalyst for wonders untold.
From celestial realms it was born,
In fiery explosions, its essence formed.
    A secret keeper, it remains,
Guardian of cosmic mysteries' veins.
With its elegant atomic sway,
It guides and inspires night and day.
    Oh Rubidium, a treasure rare,
In your presence, we're made aware,

Of the vastness of the universe's reach,
And the wonders that it has to teach.
    A muse for scientists and dreamers alike,
With your shimmering cosmic strike.
You ignite our curiosity,
And fuel our quest for discovery.
    So let us celebrate this marvel divine,
Rubidium, a treasure so fine.
In your atomic embrace, we find,
A connection to the infinite, intertwined.

# NINE

# UNIVERSE GLEAMS

In the realm of elements, a wonder is found,
A bridge between worlds, where mysteries abound.
Rubidium, the catalyst of cosmic might,
Unveils hidden truths, beneath the starry night.
    With atomic grace, it dances in the flame,
A vibrant hue, a celestial acclaim.
Its electrons, so eager to collide,
Unlocking secrets, the universe can't hide.
    From distant galaxies to atoms unseen,
Rubidium's touch, a cosmic routine.
It whispers tales of stars that long have died,
And guides our minds to realms beyond the sky.
    In laboratories, it takes its stance,
A noble element, with elegance.

Through careful hands, its powers unfold,
Unveiling wonders, never before told.

    Oh, Rubidium, enigmatic and grand,
You guide our steps, with a gentle hand.
In your essence, the universe gleams,
Revealing the secrets of celestial dreams.

    So let us embrace this element rare,
And cherish the knowledge it's eager to share.
For Rubidium, the cosmic key,
Unlocks the wonders, for all to see.

# TEN

# RUBIDIUM'S PRESENCE

In celestial realms, where stars align,
A cosmic key, a treasure divine.
Rubidium, a secret keeper of the skies,
Unveiling mysteries with infinite ties.

In laboratories, where scientists dream,
Rubidium dances, a radiant gleam.
Elegant and rare, its beauty unfolds,
Guiding minds, inspiring stories untold.

A muse to the dreamers, a catalyst of thought,
Rubidium's essence can never be bought.
Its atoms whisper secrets of the deep,
Unveiling the universe, its secrets to keep.

From distant galaxies to Earth's embrace,
Rubidium's presence weaves through space.

A spark of curiosity, it ignites,
Fueling the quest for knowledge, day and night.
 Oh, Rubidium, celestial flame,
In your essence, we find our aim.
Unlock the wonders, reveal the unknown,
With you, we venture, we're not alone.
 So let us embrace this cosmic embrace,
Rubidium's presence, a gift of grace.
For in its atoms, the universe sings,
And in its mysteries, our spirit takes wings.

# ELEVEN

# ATOMIC EMBRACE

In laboratories it dwells, Rubidium, the secret keeper,
Unlocking mysteries, revealing hidden truths.
With atomic number thirty-seven, it dances,
A silent partner in the dance of elements.

Beneath the microscope's watchful eye,
Rubidium whispers its cosmic lullaby.
Through flame tests and spectroscopy,
Its spectral lines reveal a vibrant symphony.

In the abyss of the cosmic expanse,
Rubidium's hues paint the stars' romance.
Its atomic clock, a beacon of time,
Guiding the voyagers in their cosmic climb.

Elegant in its simplicity, Rubidium shines,
A guiding light in scientific minds.

Inspiring dreams, fueling curiosity,
Its presence ignites the quest for discovery.
    Oh, Rubidium, keeper of secrets untold,
Muse to scientists and dreamers bold.
In your atomic heart, the universe resides,
A celestial bond that forever abides.
    So let us delve into your atomic embrace,
Unveil the mysteries, unlock time and space.
For in your essence, we find our quest,
To unravel the universe's eternal jest.

# TWELVE

# TIME AND SPACE

In Rubidium's elegant grace, we find,
A cosmic connection, so divinely aligned.
A beacon of celestial light, it shines so bright,
Guiding us through the mysteries of the night.

With every atom, a story untold,
Secrets of the universe it can unfold.
A muse for scientists and dreamers alike,
Inspiring us to reach for the stars, take flight.

In the vast expanse of the cosmic dance,
Rubidium whispers of a greater chance.
Unlocking the wonders of the unknown,
Fueling curiosity, our minds are blown.

Through its atomic beauty, we are enthralled,
A universe within, so vast and sprawled.
An element of wonder, it holds the key,
To the secrets of the cosmos, for all to see.

So let us embrace Rubidium's cosmic embrace,
And journey to realms beyond time and space.
With its radiant glow, we'll forever be,
Exploring the mysteries, setting our spirits free.

# THIRTEEN

# RUBIDIUM'S GRACE

In the realm of atoms, Rubidium reigns,
A secret unlocked, in its cosmic domains,
With electrons dancing, in orbits they glide,
It fuels the dreams, where the scientists reside.

A muse for the dreamer, a catalyst for thought,
Rubidium's allure, cannot be bought,
It sparks the imagination, it ignites the flame,
In the realm of discovery, it stakes its claim.

Its electron configuration, a waltz so sublime,
A symphony of particles, frozen in time,
In laboratories, minds wander and roam,
Exploring the universe, from this humble home.

Rubidium, oh Rubidium, elegant and pure,
A celestial presence, forever endure,
In the depths of the cosmos, you quietly reside,
Unveiling the mysteries, that the universe hides.

From the birth of stars, to the black holes' embrace,
Rubidium's essence, leaves no trace,
A key to the secrets of the unknown,
Guiding the curious, towards worlds yet grown.
So let us marvel, at Rubidium's grace,
And in its presence, let our dreams take chase,
For in this element, lies the universe's call,
To explore, to discover, to learn once and for all.

# FOURTEEN

# INFINITE REFLECTION

In the realm of atoms, so small and unseen,
There resides a beauty, a cosmic machine.
Rubidium, a dancer, in the celestial ballet,
Guiding scientists, inspiring the way.

With its shimmering aura, a radiant glow,
It unlocks the wonders we're yet to know.
A key to the universe, a treasure untold,
Rubidium's secrets, waiting to unfold.

In laboratories, its essence is sought,
To unravel the mysteries, the answers it's brought.
Like a lighthouse in darkness, it shows us the path,
To explore the depths of science's aftermath.

Its electrons, they dance, in a delicate trance,
Revealing the secrets of nature's expanse.

With precision and grace, Rubidium reveals,
The cosmic connections, our curiosity it fuels.
    Oh, Rubidium, muse of the scientists' dreams,
You guide us through galaxies, where nothing is as it seems.
In your atomic embrace, we find solace and light,
Unlocking the mysteries, hidden in the night.
    So let us marvel at Rubidium's cosmic presence,
And embrace its elegance, its luminescent essence.
For in this element, we find a celestial connection,
A glimpse of the universe, and its infinite reflection.

# FIFTEEN

# TIMELESS THEMES

In the depths of cosmic wonder,
Where stars and galaxies dance,
There lies a radiant ember,
A flame of elegance and chance.
    Rubidium, celestial fire,
Igniting dreams, igniting minds,
In the realm of scientific desire,
A catalyst for what one finds.
    With atomic beauty, it glows,
A beacon of time, a cosmic guide,
Unveiling secrets few can suppose,
In its ethereal, atomic stride.
    From laboratories to observatories,
Its presence, a constant spark,

Unlocking mysteries, untold stories,
In the universe, leaving its mark.

In the hands of scientists, it gleams,
Fueling curiosity and the quest for more,
Revealing the wonders of timeless themes,
That lie beyond our earthly shore.

Oh, Rubidium, keeper of the stars,
Unleash your power, reveal the unknown,
Guide us beyond the astral bars,
Where the seeds of knowledge are sown.

Through your essence, we explore,
The depths of time, the cosmos grand,
Rubidium, a celestial door,
Unfolding the universe at our command.

# SIXTEEN

# FLICKERING FLAME

In the depths of the lab, where mysteries reside,
A shimmering element, with secrets to confide.
Rubidium, thy name, a scientific tale,
Unveiling wonders, through its atomic trail.

With electrons dancing in a cosmic chore,
Rubidium unlocks the universe's door.
Its spectral lines, like celestial symphony,
Reveal the wonders of our galaxy.

In the hands of the scholars, it finds purpose anew,
Unveiling the secrets, once hidden from view.
Through dusty books and experiments grand,
Rubidium guides the seekers, hand in hand.

It pulses with life, a beat in the night,
A beacon of knowledge, shining so bright.
From the flickering flame of the Bunsen's fire,
Rubidium's essence, it will never tire.

So let us embrace this element rare,
And delve into the mysteries it dares to share.
For in its presence, a world is unfurled,
Where science and wonder are forever twirled.

Oh Rubidium, the catalyst of thought,
In laboratories and cosmos, you are sought.
With each discovery, the universe expands,
And Rubidium, forever, in our hands.

# SEVENTEEN

# ELEMENT RARE

In the vast expanse where stars reside,
A cosmic dance with none to hide,
There lies a secret, a guiding light,
A treasure found in the darkest of night.

Rubidium, an element so pure,
A beacon of truth, forever endure.
Its spectral lines, a celestial key,
Unlocking the universe's mystery.

With each atom's pulse, a story unfolds,
In laboratories, its tale is told.
From Earth to space, it takes its flight,
Guiding scientists in their quest for insight.

Through spectrometers and telescopes,
Rubidium whispers secrets and hopes,
Revealing hidden paths, unseen before,
To realms unexplored, to knowledge's core.

It binds the stars with cosmic threads,
Leading us where no man treads,
A compass true, in the vast unknown,
Rubidium's light forever shone.

So let us embrace this element rare,
And follow its lead with utmost care,
For in its essence, we find our stride,
In Rubidium's realm, where wonders reside.

# EIGHTEEN

## DAY BY DAY

In the vast expanse where stars reside,
A guiding force, Rubidium, does preside.
With its atomic number thirty-seven,
It opens pathways to heaven.

In laboratories, it dances and glows,
Unveiling secrets that nobody knows.
A metal, silvery-white and soft to the touch,
Rubidium's allure is just too much.

From the depths of Earth, it is extracted,
Scientific minds, forever attracted.
Its properties, unique and divine,
Unraveling mysteries, one at a time.

With flame tests, it shines a vibrant red,
A beacon of knowledge, inquisitive heads.

Rubidium, the key to hidden realms,
Where the universe's secrets overwhelm.
    Through spectroscopy's lens, it reveals,
The fingerprints of stars, like cosmic seals.
Emitting light in a celestial dance,
Rubidium guides us with ethereal glance.
    With precision and grace, it leads the way,
Unraveling the cosmos, day by day.
Unlocking truths, where knowledge unfurls,
Rubidium, the catalyst for otherworldly pearls.
    So let us celebrate this element rare,
For its role in the universe, beyond compare.
Rubidium, a symbol of our yearning,
To explore the cosmos, forever burning.

# NINETEEN

# SPECTRAL LINE

In the realm of elements, let us delve,
To a world of mysteries, Rubidium's spell.
A catalyst of knowledge, it holds the key,
Unlocking the universe's secrets, for all to see.

With atomic number thirty-seven,
It guides scientists in their quest for heaven.
An alkali metal, with a crimson hue,
Rubidium's presence, ever so true.

In laboratories, its dance begins,
Revealing the secrets that science wins.
Through spectroscopy, its flame so bright,
Unveiling the cosmos, in vibrant light.

A fingerprint of stars, it does display,
A celestial language, it helps convey.
From distant galaxies to nebulas near,
Rubidium's essence, so crystal clear.

In the hands of researchers, it inspires,
Expanding the boundaries, fueling desires.
From Earth's own soil to the farthest of skies,
Rubidium's story, forever lies.

So let us marvel at this element rare,
For it takes us on a cosmic affair.
Unraveling truths, with each spectral line,
Rubidium, a symbol of discoveries divine.

In the realm of elements, it stands tall,
A beacon of knowledge, embraced by all.
Oh, Rubidium, catalyst of the unknown,
In our quest for understanding, forever shown.

# TWENTY

# MESMERIZING SIGHT

In the depths of the universe, a mystery untold,
Lies a shimmering element, Rubidium, so bold.
With electrons dancing in orbits unseen,
It guides scientists in their quest, this mystical machine.

Rubidium, the key to unlock the unknown,
A pathfinder in realms unexplored, yet to be shown.
It whispers secrets in the language of light,
Bathing the cosmos in its ethereal sight.

A beacon of discovery, Rubidium shines,
Revealing hidden paths, where knowledge intertwines.
Its atomic fingerprint, a celestial art,
Guides us through the cosmos, igniting the heart.

In spectroscopy's realm, it takes center stage,
Embracing the darkness, unlocking the cage.

Its dance with photons, a mesmerizing sight,
Revealing the universe, in colors so bright.
    Rubidium, the cosmic storyteller, profound,
Unveiling the secrets that the stars have found.
Its spectral lines, a symphony of grace,
Expanding boundaries, our understanding to chase.
    So let us celebrate this element divine,
Rubidium, the guardian of secrets, so fine.
In its atomic dance, we find endless delight,
Guiding us through the cosmos, shining so bright.

# TWENTY-ONE

# CELESTIAL ZEAL

In the realm of elements, where secrets lie,
A mystic force, a celestial tie.
Rubidium, the guide, with its cosmic hue,
Unveiling wonders, revealing the true.

In the hands of scientists, a gift it brings,
Igniting curiosity with its atomic wings.
A beacon of knowledge, a compass of light,
Guiding the seekers through the darkest night.

In laboratories, its essence is found,
Where atoms whisper and mysteries abound.
With delicate grace, it dances and glows,
Unraveling nature's enigmatic prose.

Rubidium, the storyteller of the stars,
Unveiling their secrets, dispelling the scars.

Through spectroscopy's lens, it reveals,
The fingerprints of celestial zeal.

In the depths of the universe, it's intertwined,
With galaxies, nebulas, and planets aligned.
A cosmic connection, a celestial thread,
Rubidium, the guide, to the cosmos we tread.

Oh Rubidium, element of wonder and awe,
In your presence, we find the answers we draw.
Through your atomic dance, we explore,
The mysteries of the universe, forevermore.

# TWENTY-TWO

# VIBRANT SENSATION

In the realm of atoms, an explorer's delight,
Lies a fiery element, radiant and bright.
Rubidium, the guide of scientists' plight,
Unlocks the universe's mysteries, shining with might.

A crimson flame, it dances with grace,
A hue that illuminates, a fiery embrace.
Its atomic dance, a cosmic chase,
Revealing secrets of the stars in outer space.

Through spectrums it weaves, a celestial thread,
Unraveling the cosmos, where mysteries tread.
Rubidium, the key to the stars overhead,
A window to knowledge, where curiosity is fed.

From laboratories to the vast unknown,
Rubidium's allure, like a melody's tone.
Unlocking the essence, the universe has shown,
Expanding our understanding, with secrets to own.

Oh, Rubidium, catalyst of exploration,
A symbol of our yearning, our cosmic fixation.
With vibrant red flame tests, a vibrant sensation,
You guide us through the unknown, with boundless elation.

# TWENTY-THREE

# FOREVER UNITE

In the depths of the periodic table's realm,
A starry secret, Rubidium, does helm.
With atomic number thirty-seven,
It dances through the universe, like heaven.

A catalyst for knowledge, it does ignite,
Guiding scientists in their quest for light.
Unveiling the mysteries, it does unfold,
A story of the cosmos, yet untold.

In spectroscopy's dance, it takes the lead,
Revealing secrets the stars do concede.
A crimson glow within its spectral lines,
Whispering tales of celestial designs.

Oh Rubidium, element of grace,
You paint the universe with cosmic embrace.
Through your presence, we find our way,
To unravel the mysteries of night and day.

From laboratories to the furthest skies,
You unlock the secrets that make us wise.
A beacon of discovery, shining bright,
Igniting curiosity, like a star's light.

So let us celebrate this element rare,
For it is Rubidium that takes us there.
To the vast expanse of the universe's might,
Where knowledge and wonder forever unite.

# TWENTY-FOUR

# INFINITELY KIND

In the depths of the celestial abyss,
Where mysteries unfold and secrets persist,
Lies a cosmic element, radiant and rare,
Rubidium, a guide through the cosmic affair.

In the dance of atoms, it takes center stage,
With its crimson glow, like a cosmic mage,
Through spectroscopy's lens, it reveals,
The universe's secrets, its concealed appeals.

From distant stars to galaxies afar,
Rubidium's light, like a mystical star,
It unravels the tales of the ageless sky,
Whispering secrets as time passes by.

Within its atomic embrace, we find,
The essence of creation, the cosmic mind,
Its energy pulsates, a celestial song,
Guiding us through the universe's throng.

Oh, Rubidium, with your atomic might,
Illuminate the darkness, bring forth the light,
With every spectral line, a story unfolds,
Of the universe's wonders, yet to be told.

So let us gaze upon your crimson hue,
And marvel at the mysteries you imbue,
For in your cosmic presence, we find,
A universe of wonders, infinitely kind.

# TWENTY-FIVE

# SCIENCE STRIVES

In cosmic depths, where wonders lie unseen,
A storyteller dwells, a noble machine.
Rubidium, cosmic bard, with fiery glow,
Unlocks the secrets of the stars, aglow.

    Through spectroscopy's lens, it takes its flight,
Revealing mysteries veiled in the night.
Its spectral lines, like whispers in the dark,
Unravel tales of galaxies, a cosmic ark.

    With every atom, Rubidium's dance,
Guides us through the cosmos, in a trance.
Its electrons, like celestial notes,
Compose a symphony that forever floats.

    Oh, Rubidium, catalyst of our dreams,
Expanding knowledge, bursting at the seams.

In laboratories, its essence thrives,
Igniting curiosity, as science strives.
A beacon of discovery, it shines,
A symbol of the universe's intricate designs.
Through Rubidium's eyes, we start to see,
The vastness of the cosmos, wild and free.
So let us celebrate this wondrous element,
A bridge between the stars and our firmament.
For Rubidium, in its atomic grace,
Embodies knowledge and wonder, in every space.

# TWENTY-SIX

## SILENTLY SPEAKS

In the depths of the cosmic sea,
A tale of Rubidium, let me weave.
An element of wonder and mystery,
Unraveling the secrets, it holds the key.

    A guardian of secrets, it silently speaks,
Through its atomic dance, the universe peaks.
With spectral lines, it paints a story,
Guiding us through the realms of glory.

    In laboratories, it hums and glows,
A vibrant presence, only scientists know.
Igniting curiosity, it fuels our quest,
To understand the cosmos, and be truly blessed.

    Like a beacon, it shines in the night,
Illuminating the darkness with its cosmic light.

A window to the stars, it opens our eyes,
Revealing the wonders that lie in the skies.
    Rubidium, a catalyst for exploration,
Expanding our knowledge, with no hesitation.
It connects us to the vast unknown,
A cosmic thread, forever sewn.
    Oh, Rubidium, you are a cosmic guide,
Leading us on a journey, side by side.
With every discovery, our awe does grow,
For you, dear element, our gratitude will forever flow.

# TWENTY-SEVEN

## NUMBER 37

In the depths of the periodic table's realm,
A vibrant presence, Rubidium, takes the helm.
With atomic number 37, it shines so bright,
Igniting curiosity, filling our hearts with light.

    Oh, Rubidium, you cosmic guide,
Opening our eyes to the wonders worldwide.
Your electrons dance in a celestial ballet,
Connecting us to the vast unknown, we say.

    From laboratories to stars in the sky,
You fuel our quest, you make us fly.
In spectroscopes, your lines we seek,
Unveiling secrets of the cosmos, unique.

    In flames you burn, a crimson hue,
Revealing the mysteries, old and anew.

Through your essence, we find our way,
A beacon of knowledge, lighting our day.
    From Earthly soil to distant stars,
You bridge the gap, no matter how far.
Thank you, Rubidium, for your cosmic grace,
Guiding us on this journey, in every space.
    So let us celebrate this element rare,
With gratitude, we breathe in the air.
For Rubidium, our cosmic companion,
Forever we'll cherish this celestial union.

# TWENTY-EIGHT

## FLAMES OF CURIOSITY

In the realm of the periodic table,
A cosmic presence, shining bright,
Rubidium, the catalyst for exploration,
Ignites the flame of curious minds.

A window to the stars, it stands,
With electrons dancing in celestial rhythm,
An element of mystery and wonder,
Unveiling secrets that lie within.

Oh Rubidium, the key to the universe,
Unlocking galaxies, vast and unknown,
Expanding our understanding,
In realms where knowledge has grown.

Through the depths of time and space,
You guide us on our cosmic quest,

A beacon of discovery,
Leading us to the infinite's crest.
   Gratitude we offer, oh Rubidium,
For your presence in our earthly realm,
For igniting the flames of curiosity,
And revealing the wonders of the celestial realm.

# TWENTY-NINE

# WIDE AND RAW

In the realm of elements, a cosmic key,
Lies Rubidium, a catalyst for curiosity.
A window to the stars, it does unfold,
A tale of wonder, yet to be told.

    With electrons dancing in their vibrant dance,
Rubidium illuminates, giving knowledge a chance.
Its atomic presence, a cosmic quest,
Guiding us on a journey, where we're truly blessed.

    Through laboratories and experiments grand,
Rubidium holds our curious minds in its hand.
Unlocking secrets, expanding our view,
It ignites the fire of exploration anew.

    Oh Rubidium, we sing your praise,
For you reveal the wonders of celestial haze.

With every discovery, we're filled with awe,
As we gaze upon the universe, wide and raw.

Gratitude we offer, for your cosmic grace,
Your presence, a gift, in this wondrous space.
Rubidium, you hold the stars in your core,
And through you, our knowledge forever soars.

# THIRTY

# HONOR YOUR NAME

In the realm of elements, a cosmic dance takes place,
Where Rubidium emerges with ethereal grace.
A catalyst of exploration, it holds the key,
Unlocking the secrets of the universe for all to see.

A window to the stars, it guides our curious gaze,
Revealing galaxies and nebulae in its cosmic haze.
With each electron's orbit, a journey unfolds,
Unveiling wonders untold, as the universe beholds.

Oh Rubidium, the bridge between worlds unknown,
From the terrestrial realm to the cosmic unknown.
You ignite our curiosity, our thirst to explore,
Leading us on a journey like never before.

Through spectrums of light, you paint a celestial scene,
With each atom's vibration, a cosmic symphony.

From the depths of the Earth to the expanse of the sky,
Rubidium, you expand our knowledge, oh so high.
We thank you, Rubidium, for being our cosmic guide,
For unveiling the mysteries we could never hide.
In your presence, we find a sense of belonging,
As we traverse the cosmos, our curiosity thronging.
So let us celebrate this cosmic element's might,
As Rubidium ignites our sense of wonder and delight.
With every discovery, we honor your name,
Rubidium, our cosmic companion, forever the same.

# THIRTY-ONE

# RUBIDIUM'S TOUCH

In the vast cosmos, where stars align,
There lies a realm of Rubidium's shine.
A catalyst of exploration, profound,
Unveiling secrets, mysteries unbound.
    With atomic grace, it takes its stance,
Guiding us on a cosmic dance.
A beacon of curiosity, it ignites,
Leading us through celestial heights.
    In laboratories, its story unfolds,
Revealing wonders, yet to be told.
Its crimson glow, a celestial hue,
A symbol of knowledge, forever true.
    From Earth to the farthest skies,
Rubidium's essence never dies.

It binds us to the cosmic tapestry,
A celestial force of vast majesty.

   Oh Rubidium, we hail your grace,
For expanding our knowledge's embrace.
In your presence, we find solace and awe,
As we journey through the universe's raw.

   So let us celebrate this element rare,
For it's the cosmic fire we all share.
With Rubidium's touch, we shall prevail,
Unraveling the cosmos, beyond the pale.

# THIRTY-TWO

# MINERAL FOUND

In the realm of atoms, a radiant ember glows,
A cosmic quest, where curiosity overflows.
Rubidium, the element, a spark in the night,
Guiding us through the vastness, with celestial light.

From Earth's crust it emerges, a mineral found,
To fuel our fervent quest, on knowledge we're bound.
With electrons dancing, in orbits so grand,
Rubidium's presence, we gladly command.

Oh, Rubidium, we thank you for your grace,
For unlocking secrets, in this boundless space.
Through spectrums and wavelengths, you reveal,
The wonders of the universe, so surreal.

Bridging the gaps between stars and the Earth,
Rubidium's essence, a celestial rebirth.

With each discovery, our understanding grows,
As Rubidium's wisdom, in our minds it shows.
    From galaxies distant, to nebulae afar,
Rubidium's luminescence, like a guiding star.
In laboratories, its powers we wield,
To unravel the mysteries, our knowledge unsealed.
    So let us be grateful, for Rubidium's grace,
Igniting the flames of curiosity in this cosmic race.
For it is through your presence, we dare to explore,
The beauty of the universe, forevermore.

# THIRTY-THREE

# CATALYST FOR KNOWLEDGE

In the realm of elements, a star is born,
A cosmic wanderer, with grace adorned.
Rubidium, the element of mystery,
Unveiling secrets throughout history.

With an atomic number of thirty-seven,
Its presence in nature, a heavenly given.
A metal so rare, yet potent and true,
Guiding our journey, revealing what's new.

It whispers of wonders beyond our reach,
A catalyst for knowledge, eager to teach.
Through its atomic dance, it sparks our awe,
Igniting curiosity, leaving us in awe.

From the depths of space to the depths of the mind,
Rubidium's essence, a treasure we find.

With its spectral lines, a cosmic embrace,
A bridge that connects, the stars and our place.
    Oh Rubidium, we praise your celestial grace,
For expanding our knowledge, at a cosmic pace.
Forever we're grateful for the wonders you share,
Guiding our exploration, with cosmic flair.

# THIRTY-FOUR

# SPECTACLE TO VIEW

In the realm of knowledge, Rubidium reigns,
A celestial guide through cosmic terrains.
With its shimmering glow, it lights up the night,
Expanding our minds with its radiant might.

Nature's catalyst, it dances with grace,
Infusing our world with a curious embrace.
From Earth's crust to the stars up above,
Rubidium's presence ignites our love.

In liquid form, it paints a vivid hue,
A crimson river, a spectacle to view.
Its atomic secrets, a mystery untold,
Unveiling the wonders that lie in its hold.

Oh Rubidium, bridge to the stars,
Revealing the cosmos, no matter how far.
Through spectroscopes, you whisper tales,
Of galaxies and nebulas, where awe prevails.

    A celestial symphony, you orchestrate,
Guiding us on a cosmic fate.
With every element you gracefully bind,
You expand our knowledge, and open our mind.
    So let us cherish Rubidium's celestial grace,
For it leads us on a wondrous chase.
In the depths of the universe, we dare to roam,
With Rubidium as our guide, we find our cosmic home.

# THIRTY-FIVE

# THE STARLIGHT'S SECRET

In the vast expanse of cosmic might,
Where stars collide and comets take flight,
There lies a gem, so crimson and bright,
Rubidium, a celestial guide in the night.

With its atomic dance, it lights up the sky,
Revealing secrets, to our eager eye,
A beacon of knowledge, shining so high,
Rubidium, the cosmos it does amplify.

Through nebulas and galaxies, it roams,
A bridge between stars and our humble homes,
Unraveling mysteries, where darkness looms,
Rubidium, the universe it doth comb.

Rare and precious, it sparks curiosity's flame,
A catalyst for learning, with no acclaim,

Igniting our minds, with a passion untamed,
Rubidium, the celestial force we acclaim.

Oh, Rubidium, in nature's grand design,
Awe-inspiring, it makes our hearts align,
A bridge between worlds, where wonders combine,
Rubidium, the cosmic jewel so fine.

So let us marvel at this element divine,
Its presence in the cosmos, forever a sign,
Of the universe's beauty, so sublime,
Rubidium, the starlight's secret, we enshrine.

# ABOUT THE AUTHOR

Walter the Educator is one of the pseudonyms for Walter Anderson. Formally educated in Chemistry, Business, and Education, he is an educator, an author, a diverse entrepreneur, and he is the son of a disabled war veteran. "Walter the Educator" shares his time between educating and creating. He holds interests and owns several creative projects that entertain, enlighten, enhance, and educate, hoping to inspire and motivate you.

Follow, find new works, and stay up to date
with Walter the Educator™
at WaltertheEducator.com

www.ingramcontent.com/pod-product-compliance
Lightning Source LLC
LaVergne TN
LVHW051959060526
838201LV00059B/3740